DC LANFRANCHI

LEADERSHIP IN SAFETY

THE RED AND WHITE
BOOK OF HSE

DC Lanfranchi

LEADERSHIP IN SAFETY

THE RED AND WHITE BOOK OF HSE

Copyright 2019

www.resiliere.com

All rights reserved.
Published by Deborah Lanfranchi
Amazon Kindle Direct Publishing

15746 words.

First Edition
September 2019
ISBN 9781692022778

To my family and to all my employees, seniors and organizations from whom I have learnt and with whom I shared my passion and mission, and to their families as well.

INDEX

INDEX 7

PROLOGUE 9

PREFACE 11

DIFFERENCE BETWEEN *SAFETY* & SECURITY 12
WHAT DOES IT MEAN TO BE A GOOD LEADER? 13
HOW TO TRAIN LEADERSHIP? 13
WHICH IS MY OBJECTIVE? 14
FROM RED TO WHITE 17

INTRODUCTION 23

PART 1 ANTICIPATING 29

HSE – THE IMPORTANCE OF SAFETY 31
HSE – THE ORIGIN OF SAFETY 32
   Modern Concept of Safety   35
LEADERSHIP IN SAFETY 37
   Second competence: Coherence   42
TEAMS MANAGEMENT - THE IMPORTANCE OF BELIEFS. 45
   Limiting Beliefs   46
FINDING AN HSE LEADER 47
   Is leadership won or acquired?   49
   The leader is the team´s mirror   50
HOW TO BECOME A SAFETY LEADER 51
   Organizational Change and its Developing Phases   52
   Kairos Moment   53
   Common sense   54

PART 2 TAKING ACTION 57

ACTION AXES 59
   Communication   59
   Keys to effective communication   61
   Communication and Cultural Differences   63
   Communication and Cognitive styles   63
NEGOTIATING STRATEGIES 64
   Faith and Evidence   64

Systems of Values	65

PART 3 MEASURING 69

THE IMPORTANCE OF COMBINED VARIABLES 71
    Comparing Progress   73
    Seniority of Reports   73
DETECTION OF MATHEMATICAL PATTERNS.................................... 76
HOW IMPORTANT IS IT FOR A LEADER TO MEASURE? 79

PART 4: AFTERTHOUGHTS & REFLEXIONS 81

WHY DO CRISES HAPPEN AND THE END? APRIL 2020 REFLECTIONS.. 82
ERGONOMICS.. 86
BUILDING HAPPINESS 90
HEALTHY ROUTINES 91
STRESS.. 95

EPILOGUE.. 97

DEBORAH C. LANFRANCHI 99

EDUCATION 99

Prologue

As a manager in a multinational company for so many years, meeting Deborah and getting to know her vision has been both enriching and challenging.

Deborah has an undisputable commitment with *Safety practices* and is one of highest qualified professionals I have met in the industry. Her expansive energy and undeniable values have offered a safe environment for many working habitats and to thousands of workers.

When she told me she was writing a collection of books (yes, a collection!), I was not surprised at all. Her knowledge is vast and needs to be divided into segments. Shortly afterwards, Deborah showed a preliminary document focused on leadership and I considered it marvelous; easy-reading but deep, passionate but professional, serious but entertaining. She commented she was working with writers and consultants on the contents of the following books, including matters such as Change Management, Rules and Procedures. All of these matters contribute with the different RSE divisions.

Her enthusiasm encouraged me and her passion, knowledge and commitment captivated me. The leader must explain, inspire and set an example. All these actions were transmitted to me by this work. This is why I wanted to write this prologue, brief and straightforward. I thus invite you to feel enthusiastic too, to set aside any kind of exhaustion and to enjoy Deborah C. Lanfranchi´s contagious energy. I want to emphasize her contribution with a clear methodology inspired in Dupont´s Bradley curve which she called PAM: **Anticipating, Taking Action and Measuring.** (ATM).

I personally hope that this collection of books will soon become teaching material for company workshops at all levels. I wish that Deborah instills her passion, expertise and commitment into those responsible for safety and that she inspires them to imitate Bradley´s spirit and hers and thus to make it possible for safety leaders to develop accurate, clear, easy procedures, and mainly to be able to sustain them in time. I expect these lines help the reader to engage in DC Lanfranchi´s methodology. They are, undoubtedly, the key to develop and improve workers ´safety, contributing at the same time to their motivation, efficiency and responsibility.

Thank you Deborah for this opportunity to support you and for the chance you offer me to join your mission with passion, knowledge and commitment.

Gabriel Rodríguez Garrido

September 2019

Preface

Why have I decided to write books? Because through more than 20 years of experience in **HSE** positions (Health *Safety* and Environment), I encountered very poor safety knowledge and practices and I immediately understood that the consequences of this institutional poverty were fatal, literally fatal. This led me to become aware that the concept of safety should be reviewed in depth.

In my first professional years I frequently heard that the real safety leaders are the organization leaders, while the experts are consultants. After hearing and experiencing it so many times I eventually convinced myself that this concept was right and it was thus meant to be.

My awakening came when I started working for DuPont as *Safety Consultant*. It was there that I clearly understood that safety is nothing else than **the protection of people´s lives**. These people might be working mates, university colleagues and even friends.

This helped me understand the importance of the impact safety has on every area of the business, on people´s perception that they ARE and FEEL safe. This importance is undeniably the foundation of everything, the structure of everything. Consequently, I clearly saw that achieving to make people be and feel safe could not be just a priority.

Stating that safety is a priority would mean having to put something before something else or placing safety before other matters. Offering a safe environment is a requirement not a priority. Safety is a basic need because it is the minimum requirement to conduct work efficiently and effectively.

Being safe must be an immutable VALUE in the organization and a quality shared by all. But watch out, on many occasions when we state that safety is the responsibility of everyone (a message present in every *Safety* boards), it ends up being nobody´s responsibility. It is highly important to feel deeply in our hearts the concept that everybody is responsible for safety. And that it can only be achieved following accurate and clear steps that require great commitment and leadership.

Difference between *Safety* & Security

Both words Safety and Security in Spanish refer to "Seguridad", but in English they mean different things. This difference lies in the origin of this safety, that is to say whether safety is deliberate or not.

A safe environment is the one where we anticipate to the occurrence of accidents that are the result of non deliberate coincidences.

Meanwhile, a safe or SECURE environment is the one that ensures protection against deliberate events that go against order, peace, safety and integrity.

I then use the word "SEGURIDAD", including in it both connotations of the English terms SAFE and *SAFETY*. Both words aim to achieve an environment where work is safe and efficient. This safety is neither acephalous nor self-imposed. On the contrary, Safety can only be conducted successfully, efficiently and effectively by a good leader.

What does it mean to be a good leader?

This is a fascinating subject that I explain in detail and easily in the following pages. But to introduce the subject I would like to say something apparently easy but not always put into practice. The real safety leaders are the leaders of an organization. They are the ones who conduct people the ones who make things happen. In my opinion, this concept is utterly clear, but to my surprise it seems not to be that clear in organizations.

The question "What is the role of those in charge of safety if I am also asked to be responsible for it?" this was and still is an everyday dilemma. I may then conclude that it is necessary to start from scratch explaining the concept of leadership.

How to train leadership?

How to do this without going back to the huge old narrative of real referents of this subject?

Luckily, I easily found the answer. It is not necessary to speak in a difficult language, we have to be deep not only theoretical, we have to share knowledge, skills, and experience and above all, it is necessary to be clear, straightforward and accurate. It is fundamental to provide uncomplicated and effective rules and procedures which would generate habits.

That is the key: that all our habits are supported by our beliefs and likewise that our beliefs are supported by our habits. The challenge is not only to explain, prove and train, but also to sustain it in time. To change bad habits for good ones and to add new good habits each time.

SUSTAIN = DEVELOP HABITS

Safety habits and Leadership habits.

When I take action, I reason, plan and execute it and then I assess the consequences. After repeating this action several times in perfect order, I turn them into a habit. Habits are incorporated to our system, to our body, to our behavior. Now we do not waste time deciding on a course of action, it is already incorporated in our operative system.

Good habits create efficiency and people with good habits create successful organizations. Instead, bad habits bring about trouble and trouble causes loss. Not only loss of profit but loss of human life as well.

Which is my objective?

My objective is to awake the holy fire every leader must possess. The real leader does not guide, is followed. I asked myself: Who would I follow? And I found an immediate answer. I would follow someone who takes care of me, who teaches me, trains me and protects me, but above all someone who encourages me to grow and who respects me fully.

Respect for human dignity within the working place is related to people´s safety conditions. If people cannot work safely, they cannot work with dignity.

Is it easy to be a HSE leader? No, it isn´t.

Why not?, because it implies to be upright and this sometimes entails to change the existing set of values to make hard decisions that at first instance seem to be harmful for the business, but which are really not. Each time that, as Production Manager, I had to make one of these decisions, I now see from a distance, they were the right decisions both at a humane level and at business level as well.

Integrity is the base of a successful business.

The objective of this book is to raise awareness of the importance of leadership in every cultural change. The leader is the mirror in which people reflect themselves, and as such has to give back an upright image. The leader is the visible face of the organization. The responsibility for their people cannot be delegated.

I would like to show this parallelism: when your children are ill you take them to the doctor. The doctor gives them some medicine and tells you how to administer such medication. Now, the following questions arise: Who is responsible for your children´s health? Who is responsible for giving them their medicine as indicated by the doctor? Who is responsible for the children´s safety? The same dilemma is faced by HSE specialists and the line leader.

I intend this collection of books to be both inspirational as well as a guidance to awake the HSE leader every organization referent should have in his inner self.

Who can disagree on protecting their people? Don´t organizations always say that their most valuable asset is their human capital? Consequently it is their unavoidable duty to protect them.

After working in HSE and manufacturing for more than twenty years I came across the word leadership thousands of times and in different contexts.

Which is the difference between a leader and an HSE leader? Basically none. They both have to protect people and the organization. Their minds, heads, principles, values and daily actions must only focus on the safety of their employees and always act accordingly (coherence).

That is to say, the difference is only centered in the type of beliefs according to which each individual acts. In other words, it is related to the mental model on which every individual builds itself. It is so easy (and difficult at the same time) as to believe that every accident is predictable and consequently I will act according to this belief, I will do my best and even more, at all times, at every place, to prevent any accident.

This book is based on taking action and explaining HSE and ends suggesting how to work in order to have a calm and healthy leader.

I will include several examples in the books and I will share this story in this preface.

> *In a plant there was a very efficient manager, he got to work before anybody else and he left after everybody else, he used to worry about every problem and always found a solution to them. People felt safe with him.*
>
> *One day, this manager died of a heart attack due to the pressure he had been subject to (I wrote heart attack, not accident, deliberately, afterwards, we will see why). People got to know the names of his relatives and sent their condolences and expressed their respect for this man. They even placed a commemorative plaque in his honor in the plant and remembered him in anniversaries. Years went by, his colleagues and the company forgot him. But his family always remembered and missed him.*

Why have I chosen this narrative? To show that to be effective we should base our course of action on the safety and health of every person, including their families. Only like this will we be really humane and humanize the others. It is very important not to make our blood boil (words that indicate stress) since this manager´s death could have been prevented.

From red to white

There is a lot of written material on HSE, EHS, SHE, or various acronyms related to **Health *Safety* & Environment**, or Safety and Hygiene, or Environment, Safety and Health. There are also loads of material available and a great amount of authors on Leadership. I do not intend to add just another book to this collection. I want to focus on the S of Safety.

HSE means much more then what we imagine and I look forward to sharing its implications and depth in different texts. I have a lot of experience as safety manager and as plant manager having worked for wonderful multinational companies and I would like to share my knowledge with you. I have planned to divide the vast material in a collection of books to make the reading and searching of information easier. I want HSE to be easy. I will start by giving each book a different color.

This year 2019 I introduced myself in the world of editors, writers, consultants, trips abroad to attend seminars and lectures where the most important HSE leaders meet. I also included in my agenda visits to plants and companies in various continents. I sincerely think this collection of books is a must. I would have liked to have it a long time ago. It would have certainly avoided a lot of suffering.

This devotion for safety did not start in my adulthood but when I was a little girl as I wanted to be an astronaut. I remember my endless curiosity about procedures, for example, I was fascinated to imagine all the parts and tools that were invented to send a spacecraft to the Moon. I also recall a recurring feeling of insecurity about being completely alone and so far away in space. This image sent shivers down my spine, from then on I concentrated my attention on this. Over the years I understood what all that meant to me. I discovered how essential it is for a human being to feel safe. We need to feel safe on every occasion, at home, at work, with our relationships, in our community, on Earth and in our future and undoubtedly also in space.

I carefully chose the colors of each book in this collection called HSE Books by DC Lanfranchi. Each book will deal with a different matter in depth which will be also found as a reference in the other books for they are related.

When Reading the yellow book on HSE values, the reader will be introduced in the world of Corporate Social Responsibility (RSE or CSR) and we will see the metrics that are relevant for our organizations.

When you read the **white** book of Change Management I will justify why external advisors are needed to introduce changes within an organization. Changing needs a lot of accuracy and objectivity and to underestimate this means to miss the opportunity to achieve optimal performance. Let´s exemplify this with a tennis match. If a tennis player has to deal with every single matter, he will not be able to play the match efficiently. Change management cannot be carried out within his working team because the perspective is an internal one and should then be provided from the outside, with an objective and highly trained 360 degrees vision of the 360 degrees comprised in HSE.

In the **light-blue** book I will talk about the future. Of the deep impact Dupont has caused thanks to his division of Sustainable Solutions (DSS – Dupont Sustainable Solutions) and of Ray Kurzweil, who has always anticipated to events. His spirit expands from his books to his leadership and his vision. In the year 2007 he founded

Singularity University, together with some other great worldwide known players in innovation.

The future will be undoubtedly molded by education. This is why I think that it is impossible to talk about the future or change management without mentioning Singularity. Larry Page, Google CEO, already in the year 2008 emphasized the importance of training to change the world. At present, Globant another highly successful multinational company at the forefront of innovation, has joined Singularity University to lead the path to the future. We´d better listen to those who know instead of trying to re-invent the wheel all the time. It will save years of bad experiences.

When we enter our destination in Google Mapsen on our telephones, the application offers different alternatives to reach our destination. Then the destination of these books is to offer alternatives so that HSE proves to be effective. Of all the different alternatives to reach my destination, I have chosen the leader path first.

As I have already said, each book will be represented by a color that varies from red to white. I want to start with a color that carries a lot of meaning, red and to finish with the color that represents the light of the sun, white.

This first book deals with the HSE leader and its symbol is the red color because this color symbolizes action, passion, stopping in front of the traffic-light or machine. It also stands for blood, real or imaginary blood. Massachusetts Technical Institute uses red and grey as its emblem and students say it represents the blood they shed on the cement to get a degree. Red also represents effort and danger.

The book of the leader should serve as a guide for anyone, from low to high positions, leaders, peers, executives, owners and shareholders. Interdependent safety is team safety. For this to be possible, it is necessary to keep in our eyes the thousands of looks we interchange in life and to understand it is the leader´s responsibility to ensure that all these people get home safe and sound.

White symbolizes purity, what is simple, clear. We should move from red to white. Nobody likes red buttons, they represent alert, risk, we stop before them with respect, but we do not pay attention to white buttons which bring us calm and peace.

I want you to work with passion (red) but to stay calm and safe (white).

I suggest incorporating a white button which reads *SAFETY* to a permanent survey. If placed strategically at certain places, the mere act of pressing it indicates that I feel safe. If not pressed, it will indicate that I do not want to participate in this survey permanently. This shows that the *Safety* parameter is not present in those who would not press it.

I invite you to press the white button to start.

Introduction

Safety is my life. I want you to feel what I feel when I speak about safety and I want you to have the best information at hand. Whether you are a young plant manager, or an experienced executive, I want you to find in my books the ticket to have access to a world of challenge and success. As I once heard, to have a program of clear, effective and efficient safety policies and to carry them out with good leadership is the ticket to play, that is to say, a minimum and fundamental requirement to play in the league of successful organizations.

I divided HSE leadership in three sections which I called ATM, (PAM). **Anticipating, taking action and measuring.** Everything is included in these three verbs. Each of them includes to define scopes, control, manage, start, finish, improve and hundreds of actions more. We must remain simple: we anticipate, we take action and we measure. I do not want to use nouns or adjectives, I want to use verbs which represent an invitation to act.

Let´s be clear and choose to use accurate language; on many occasions safety can be excessive and bureaucratic, or on the contrary, it may not gather the minimum requirements and thus may not be registered. Generally, the leader or manager may consider precautions as a rigid situation or an unimportant formality.

There are many leaders, based on the classification of Dupont's Bradley curve, based in turn on Stephen Covey efficiency principles

Following these concepts, leaders can be classified into reactive leaders, dependent leaders, independent leaders and interdependent leaders.

To put it clearly, when we describe the different types of leaders we are referring to individuals who are responsible for divisions and individuals who act differently:

- Reactive Leaders: they can only react in the face of the present circumstances.
- Dependent Leaders: They depend on what they are told.
- Independent Leaders: they are self-sufficient to take decisions by themselves.
- Interdependent Leaders: These are the leaders I really want everybody to be. Team players.

It seems to be easy, but it is not really easy at all. People must learn to play better in a team. Most of the time, people care for themselves more than for others, and this clearly brings about mistakes and mistakes cause accidents and accidents produce fatalities.

NATIONAL CENSUS OF FATAL OCCUPATIONAL INJURIES IN 2017

There were a total of 5,147 fatal work injuries recorded in the United States in 2017, down slightly from the 5,190 fatal injuries reported in 2016, the U.S. Bureau of Labor Statistics reported today. (See chart 1.) The fatal injury rate decreased to 3.5 per 100,000 full-time equivalent (FTE) workers from 3.6 in 2016. (See table 1.)

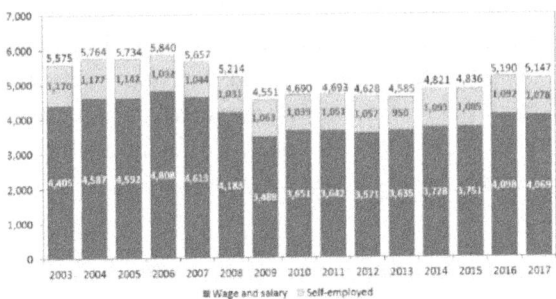

Chart 1. Number of fatal work injuries by employee status, 2003–17

According to the National Census of Fatal Occupational Injuries - Bureau of Labor, United States (https://www.bls.gov) we see there are approximately 4 fatal accidents every 100,000 employees.

Every time I learn of an accident which I am convinced could have been prevented, I feel an outburst of rage. This is exactly how real leaders feel. My wish is that this book, as easy and devoted to safety as it is, rekindles leaders´ commitment with HSE and engages more players to play the game. I want to work together the concepts of **ZERO ACCIDENTS AND ZERO TOLERANCE** with the utmost responsibility and seriousness. **FAILURE CANNOT BE AN ALTERNATIVE**.

I will provide you with a 360 degree vision of the leader role, its relationship with employees, peers, seniors, but with an institutional approach, considering people ´s bonds and with a look focused on the personal realm.

As you have already seen, I like relating examples with sports. On many occasions people think that tennis is an individual sport for only one player can be seen in the court. But to enable this player to have an extraordinary performance, it must have a team of people working together with him, trainers, kinesiologists, chiropractors, nutritionists, representatives, financial staff and above all, a family, all of them exceptional people who contribute to his emotional balance. Tennis, seen from this angle is a team sport, but for this team to work properly, institutions, clubs, providers and the whole ecosystem must work properly as well. **We can only see the tip of the iceberg**. The concept of the iceberg is an image always present in *Safety*.

So, who is a leader? It is the one responsible for people, activities, premises, procedures and the endless component parts of an organization. And this leader is required each day more and more skills. And these skills must be developed within a predictable environment and under unpredictable circumstances as well. Here is where leaders come into scene.

No doubt, each plant has a manual for each machine, but do we have a manual of all the circumstances that may take place? The same happens with people, we count with job descriptions, performance evaluations, training, but we hardly ever train for the unforeseen.

These words express my passion and conviction: Not always have accidents been unforeseen. We find in unpredictability the perfect excuse and so we hire insurance companies that cover the risks. But I want to express something as a LANFRANCHI LAW: **Team work reduces unpredictability to zero**. If we do not believe this, we have a limiting belief and we are on the path to failure.

Best Team Metrics =
Fewer Accidents

Metrics that evaluate the development of teams are closely related to mathematical trends for accidents. The better team evaluations are, the fewer accidents will occur.

We may always lose, but the leader must be persuasive as to the possibility to win. It is that simple. This is what I encourage in my workshops, a leader is 100% responsible, and to be 100% responsible they must think about the circumstances prior to, present at and following the occurrence.

The moment before is anticipating, the moment during occurrence is acting and the moment after it is measuring.

Let´s start then.

DC Lanfranchi

PART 1 ANTICIPATING

"INTERDEPENDENCE is a value superior to independence"

Stephen Covey, the 7 habits of efficient people

HSE – The importance of safety

The safety of employees and of assets in an organization is always a minimum requirement to be efficient. On many occasions I have emphasized that it is a priority, but this concept is wrong, it has no priority, it is not before or after anything. It is a critical requirement, fundamental, an immutable value, undisputable, a basic condition for an organization to work. It is as oxygen is to the body. It is not a priority it is fundamental.

> *I visited a plant and I observed that the Plant Manager had included a list of priorities in a board, placing safety in the 7th place.*
>
> *It even looked wrong at first sight.*
>
> *So I explained that to feel safe did not occupy the seventh or the first place because it was not a step in a plan. Safety was a value of the plan, with a critical strength because it was its base. Safety is a column in itself not one of its bricks.*
>
> *Safety is the foundation of a transverse column that crosses all the activities of the area and the business.*
>
> ***Working in a team and feeing safe are the critical factors of successful anticipation.***

It is a priority that the organization and all those who are responsible for people, assets and areas understand the importance of safety. Safety is as important for organizations as oxygen is for the body. Breathing is vital to live and to breathe properly must become a habit to live healthily.

At critical times during planning, the time for inhalations and exhalations is extended to 4 (four) seconds each. This breathing method brings about quietness and quietness is needed to make good decisions. On the other hand, at critical times when taking action, the time we have to make a decision may be reduced to a quarter or to the fortieth fraction of a second. In order to understand the different scenarios and to take the appropriate steps we must be trained.

HSE – The Origin of Safety

Through my professional career I was proud to work for Dupont, an organization whose safety practices I deeply admire. I would like to share some of its history.

> *DuPont Company was founded in 1802 and it has always been a leader in safety. Even at present safety still is a corporate core value in the company. Thomas Jefferson, the third President of the United States, had encouraged the founder of the company, E. I. du Pont (1772-1834), to manufacture gunpowder because at*

that time the United States did not have reliable manufacturers with high quality standards.

E. I. du Pont had unique qualifications due to the fact that he had been a disciple of Antoine Lavoisier´s the famous French Chemist who had also been the king´s public administrator. In fact, the manufacture of gunpowder was an inherently dangerous business and E. I. du Pont easily understood this. Elaborating on these dangers, he pointed out "we must try to understand the danger we live with".

His family´s and his own commitment with his employees´ safety were so strong that E. I. du Pont eventually lived in the plant. The original plan, in fact, included the requirement of building the director´s house "in such a position that the whole plant could be seen from his windows".

E.I. du Pont raised seven children in this house, and at least once his wife was injured and his house badly damaged. During an explosion, several members of the family died due to incidents related to his activity.

The 1815 explosion resulted in 9 fatalities which were the first casualties in the company. Another serious explosion took place in 1818 and caused 34 deaths, representing so far the worst procedural incident in the history of Dupont´s Company.

After an intensive process of restoration and studies on anticipating potential accidents he observed that in spite of the steps taken, people did not want to work at the plant because of fear from their lack of safety.

Finally, following several attempts to prove that their procedures were safe, Dupont ´s Company promised to incorporate the highest standards of safety. To such an extent that the founder himself moved within the

> *company's premises, offering as a guaranty of safety his own house together with his life and his wife's and children's lives.*
>
> *It was like this that he convinced workers to go back to work.*
>
> *The lessons learnt in the early years of the company's history are present nowadays in many elements of the present programs of procedural safety, which comprise safe working practices, training, research, pre-starting, safety reviews, response to emergencies and operative discipline.*
>
> <u>*Summarized fragment of Dupont's history.*</u>

Consequently, one way to start going through the safety path would entail to be fully aware of Dupont's maturity process.

We must anticipate, and not wait until a deadly accident takes place and turns us into grieving parents and grandparents, widows or widowers or children. Let's imagine a good life for everyone and let's work hard with actions that ensure a safe environment.

The leader, on behalf of the management, must be the one to guide towards safety.

> *No employee can enter a new or repaired mill until a member of the top management has operated it personally*
>
> *E. I. du Pont.*

Modern Concept of Safety

Since the first Industrial Revolution in the 18th century and to the present time, not only has the concept of work changed and new tasks appeared, but the scope of the concept of industrial safety has changed as well.

This evolution moved from the minimum and strictly necessary procedures to avoid the interruption of operations after a fatal or serious accident to a new stage introduced last century which obliges all companies to comply with the local legislation of the country of operations.

Sticking to compliance of the law had an impact on the organizations´ performance for investments on safety correlated with the operations legal and financial risk. If a lawsuit was lost, it was only then that rules were drawn to prevent incidents from happening again. In this way, all the necessary jurisprudence came into existence to elaborate rules, either by law or from previous claims.

As time went by, executives of the most important companies in the world (Forbes 50 y Forbes 500 and top 20 in all stock markets worldwide) started to develop leadership based on values supported by ethics and businesses. Not by one or the other, but by both simultaneously.

> *Ivan Seidenberg, who used to be CEO of the company Verizon, boasting more than 300,000 employees and branch offices in more than 100 countries, in 2002 said at a dinner that values such as Diversity were not*

> *just ethical but also represented value for the business itself. Values comprise both.*
>
> ***If work does not consider values, action is irrelevant.***

To develop work supported by values, organizations started recruiting real referents of Social Responsibility, people with influence in the university realm and in the community and especially with bonds and real communication with officials and governmental entities.

Big companies acknowledged the need to take care not only of their workers but also of the environment and the community where they are located.

At the beginning of the XXI century this concept started to expand even more and today the importance of Happy Companies is considered.

Technology is moving forward at giant steps and at a tremendous speed; changes sometimes appear within days, not after years or after hundreds of years: variations are significant and in almost every field of science and, if we cannot process them, we run the risk of being outside the game.

Why do we have to pay more attention to safety? Because we need companies and organizations to become more humane, family driven, kind and coherent. Technology and automation spread towards almost every daily activity: work, relationships, entertainment, leisure, relax, health, transport, communications, education. In every field we observe important technological improvements. For this reason a good leadership is a fundamental requirement to achieve safety. Without safety we will only have compasses which do not work properly.

Leadership in safety

But, what does it mean to be a leader? It is difficult to know because each organization, each division, each area defines it specifically.

The fact that a plant has the highest levels of safety confirms that there is a leader behind all that.

Let´s go over the actions of a leader. In a vast majority of organizational cultures safety is related to the word "cost". In focus groups already conducted, the word safety" or "to feel protected" appear within a positive and negative context. Legal proceedings and/or claims generally bring about inflection points and this in turn causes internal or external rules to settle via legislation.

When we talk about leadership in safety, we refer to technical competences (called hard competences) and to non technical ones (called soft competences) which a leader in an organization must possess to carry out its specific task successfully, including cultural changes when necessary.

These skills help evolve from being reactive to being dependent moving on to being independent and finally interdependent, as stated by Stephen Covey in his books "Habits to be efficient".

First Competence: Setting an example

The first competence a safety leader must evidence is to teach through its own exemplary behavior. Why did I choose this competence? For "setting an example" belongs to the realm of empathy.

Empathy was clearly defined by Daniel Goleman. He divided it in three types of empathy. Having **knowledge empathy** means to understand what the other knows, his perspective and point of view. **Emotional empathy** is needed to understand how the other person feels. And **Needs empathy**: what I need to understand is what the other person needs from me. If a person does not gather the three types of empathy within his personal competences, he cannot set an example because he is not synchronized. On many occasions I heard managers say they had set an example and that even when this was not reflected in their actions, they truly believed they had set an example. But as he did not understand the others, his example was not the one the others needed to consider him their leader, to trust him. Sometimes I feel that some people are talking about football rules when we are watching a basketball match.

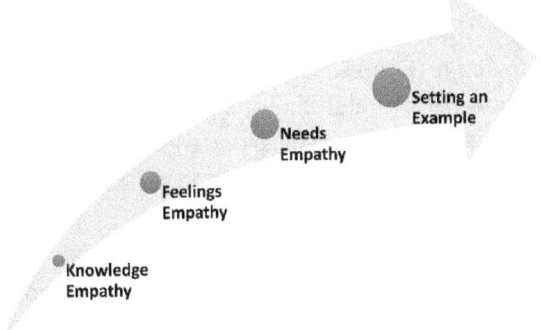

—Lanfranchi, be practical, I see that we must set an example, but how shall we start? I was very often asked.

Let´s start with **Time Management**.

If we do not administer time efficiently we make a nightmare last longer and a celebration ephemeral. We devote 80% of time to repetitive activities that cause boredom and 20% to problems which most of the times remain unsolved.

Below I include questions to see in depth the 3 types of empathies and to assess your leadership competences:

Knowledge Empathy: Do I reason what the other person knows? Do I value their point of view? Am I aware of their credentials? Are they experienced? Is the required machinery new? Was the appropriate training in procedural rules and safety rules given?

Feelings Empathy: Do I understand what the people I am in charge of feel? Do I value and respect their feelings? Do I not take into consideration those who do not think as I do? Do I know the files and personal situation of workers? Do I focus on tasks and disregard personal situations they are going through? Are they nervous? Am I calm?

Needs Empathy: Do I understand the others´ needs? Do I respect diverse opinions? Do I carry out objective assessment thoughts? or am I influenced by my personal subjectivity or the community´s subjectivity? Do I apply a correct policy of diversity and inclusion, or do I treat differently those who need something different? Do I have a 360 vision?

Now I will add some transverse questions on the subject of the different types of empathies. **Do we listen to the others? How often do they let us know, one way or the other, that they do not feel understood? Do we communicate well?**

Setting an Example: Do I carry out my activities respecting the others? Do I manage my time efficiently? Do I allocate part of my time to train and be trained? Do I use techniques which encourage mentoring among peers?

Do I encourage recognition for the others? Do I focus on the weaknesses or on the strengths of the others? Do I turn each problem into an opportunity or do I deal with it as if it were a threat?

FIRST COMPETENCE OF LEADERSHIP IN SAFETY = SETTING AN EXAMPLE WITH EMPATHY

Through example, those responsible for teams will show to their juniors the courses of action expected and accepted by the organization (since these are the ones shown by their seniors through example). Thus what is accepted at their working place will become obvious and consequently valued and recognized. Those who are responsible will also make the others open their eyes and see what is not accepted or valued. Likewise, if anybody still acted in an unacceptable way, would any kind of reprimand be expected on the part of the leader?

Many writers and leaders start by saying that leadership is based on creativity, on being proactive, on planning and on tactics and strategies. My contribution is to be practical, to support everything with examples. Setting an example is equivalent to prompting an organization into action. Some leaders will set an example using creativity, others being proactive; but setting an example will depend on the situation. It is completely different to set an example in a waste disposal plant from an example set in an administrative office.

Let´s raise a very big signal with our example and let´s be practical. Always set good examples at all realms, from life care to the management of time for different activities. We need leaders setting examples at 360 degrees.

Second competence: Coherence

We can identify leadership traits or behavior. This turns leadership into a tangible concept. Consequently, when I think of an HSE leader, I think about what I need to "see" to be ready to say this person is an HSE leader.

The first thing I need to "see" is coherence. I need to see it and to feel it, perceive it and breathe it. Generally speaking, to lead is to be a mirror. The rest of the organization reflects in the leader and when the leader is convinced of what he says, absolutely convinced of his words, this is transmitted to the rest of the organization and can be felt.

Organizational culture starts to nurture, cross purpose messages disappear and coherence develops. This is how we know what is right and what is wrong, what should be done and what should not be done. This behavior makes it easier to sustain the postulates outlined by the top management.

Is it easy to sustain coherence?

Coherence is then a course of action that can be understood through logic. Generally, we are asked to behave in a determined way, according to certain values of behavioral patterns for which we should have been prepared or trained, and what is still better, for which we should be assessed. Our performance must be measured in many areas with quantitative and qualitative parameters.

For example, we are asked to look after the people we are responsible for, but sometimes we do not support this protection with measurements. In many organizations performance is measured at the end of the year considering only the amount of boxes our team produced without taking into account if such an amount was achieved decreasing parameters such as accidents, sick leaves, behavior problems, meeting hours and several variables of quality, production and safety. And it is here where we find a course of action which lacks coherence and makes people perceive that what we always claim to be important, stops being so in favor of getting year-end bonuses.

—Lanfranchi, I see we must be coherent, how can we measure this coherence? —I often hear managers ask anxiously to go deeper into the subject.

Easy, considering that coherence is a mathematical variable which compares **combined results.** A car speed is measured in distance divided by time. Density is measured in mass / volume, there are many examples. Thus, coherence in safety implies to measure productivity divided by the safety parameters.

I share some examples of ratios that help develop coherence:

- production-profit-loss / accidents,
- production-profit-loss / sick leave,
- production-profit-loss / fatal accidents,
- production-profit-loss / lawsuits,

Even today, I still see bar, column and area graphs and I note that different measurement units are compared, or even less coherent, accidents are considered in isolation without taking into account shifts, working periods, weather conditions, or plenty of variables related to time such as seniority of secondary factors.

In the third part of this book I will include techniques to carry out really valuable measurements useful to reveal anticipatory behavior patterns. By the way, I comment here that it is very important for a good leader to be trained in Microsoft Excel.

SAFETY (A) ACCIDENTS / NUMBER OF WORKERS

SAFETY (B) = PRODUCTION /*SAFETY (A)*

Now, the way to observe if the safety trend is increasing or decreasing is evaluating this formula periodically (on a daily, weekly, monthly basis) and to calculate the <u>linear tendency</u> of the values obtained.

Microsoft Excel, dynamic table and macros operation are required to be a good leader.

Why does it seem so unnatural to be an HSE leader? Why does it seem to be so complicated? I think it has to do with this lack of consistency or lack of coherence mentioned before.

Many people in charge of teams possess a sound technical training and feel obviously at ease in these waters, but are deficiently prepared to deal with people. And, moreover, they are poorly prepared to really understand that taking care of their people is one of the duties inherent to their position. They have been promoted due to technical merit, even so, some managers conclude: *"now we´ve lost a good plant engineer and won a bad people manager"*.

Teams Management - The importance of beliefs.

A good way to conduct people teams is to use the same principles implemented to operate machinery. Both machines and people have a specific working principle. Consequently, I believe it is completely necessary for managers to understand the basic working system of human beings and likewise of people teams.

Equipment working is transverse to three axes: thoughts axis, actions axis and decisions axis.

In the thoughts axis we can find beliefs, both boosting and limiting beliefs, being the latter the ones we need to change to become true leaders. As an example and to reinforce the objective "zero accidents" we have to exercise the boosting belief that this objective is perfectly attainable. If we do not, we will never achieve it.

If in our role of safety leaders we organize our brain based on the belief that it is not possible to prevent accidents, our thoughts will use this idea to justify failure and we will act accordingly. Finally, we create what we believe in.

Action and decision axes will be dealt with in part 2 of our book: TAKING ACTION.

Limiting Beliefs

A limiting belief about HSE is the one that interprets HSE from an accounting view, and it is then considered an expense by many companies. It is like this because evidently we are looking at only one face of the coin of safety culture. It is necessary to teach that safety results in profit. I will include a personal example.

In 2006/07 a new production line was needed at the factory where I worked.

As in every new confidential Project, the negotiating table was very small and HSE was not invited to participate.

When the moment came to communicate the strategy and invite HSE to participate, the Project had already been estimated and the Money required from the head office.

Which was the problem then? Safety recommendations that would increase the variables P/L profit and loss projection had not been taken into consideration.

After we submitted our recommendations, and as we suspected, the company had to ask for more Money. Profit and loss reports could have been more accurate. Once again, both, profit and loss.

These situations illustrate the limiting belief of many companies and organizations that still consider HSE an extra cost or expense.

DC LANFRANCHI

> *HSE should be consulted, exactly as the financial department, from the very beginning. This is one of the fundamental requirements for a company to succeed.*
>
> *HSE must be a partner at the negotiating table, a Business Partner. This results in savings and benefits. Both.*

What must leaders do to become HSE leaders? Firstly they have to change their beliefs system. They have to start understanding, feeling and incorporating to their lives safety as a value.

> *Start with little things, but visible things. For example, always use the banister at the stairs. But it has to be used ALWAYS, not only when in public.*
>
> *The values that guide our lives are translated to the things we do every day, even when we are not being watched.*

Finding an HSE leader

Given that everything seems to be so easy, why is it so difficult to find out that one day you only depend on your decisions and overnight your life will depend on somebody else's decisions. Again if everything is so simple, why is it so hard to find HSE leaders or, even harder, to turn a leader into a HSE leader?

Because basically it means to modify all the belief paradigms that inhabit such a leader and also because without conscientious and hard work this is impossible.

Firstly this individual has to be willing to undergo this change, that is to say that he has to take that decision and then define the behavior that will help him rewire his habits and eventually become a HSE leader.

In other words, we need to undergo a very deep change at a personal level by which the leader's values, principles and beliefs will be modified. In fact, it all springs from a deep and true decision the leader has to take. Here is where the leader's courage appears. Or not. It is this courage the one that is essential to gather a team.

Let's be practical

LET'S WORK AS A TEAM

Which would be a good example of team work? The one in which even in the face of human life risk, people still support one another.

I would like to ask you: Can you remember the names of all the employees who work at the same building where you work, the names of their relatives and the circumstances around them? Because one day their lives might depend on your decisions, and another your life might depend on theirs.

Is leadership won or acquired?

It is very common to find expressions such as: *one is born a leader, leadership is won over, leaders are innate*, and so many more phrases describing qualities. Up to what extent are these expressions true?

On many occasions, charisma and personality traits (introvert/extrovert) are confused with leadership capability. There might be more or less charismatic leaders, but leadership results from the action of leading people and areas; and this means to conduct people and work in teams, not to be in a higher position to give orders. All of us can be leaders if we train ourselves.

People think that the path to grow into a leader might only be followed in a vertical line, that is to say climbing up the ladder, heading for managerial positions. We forget that there are other two paths towards growth.

We can also grow moving in a horizontal line, in which case the worker evolves into an expert or advisor. These are workers who do not feel at ease administering people but who are keen on growing, doing post-graduate courses; they are referents in different industries and are invited by universities. Many of them have variable salaries sometimes even higher than executives´ salaries.

The third way to grow is in a diagonal line. This path includes workers who grow vertically and horizontally. In general this group comprises workers who are transferred from one plant to the other and are part of mentors´ programs. I have met mentors who at specific times guided the acts of those above them. Being a mentor is a two way activity, as regards some matters the mentor is guided by his subordinate who happens to be an expert in a specific field and alternatively, the mentor guides his mentee to grow into a leader so that one day he could continue his task.

The above was stated by David A. Thomas, leader in Organizational Behaviors at Harvard University in his book Breaking Through, diagonal growth is the key to develop.

The leader is the team's mirror

I have mentioned before that the leader is the mirror in which everybody should wish to be reflected. Leadership is won over, is not obtained by hierarchy or inheritance. Directive positions are given by hierarchy, a leader on the contrary is followed, and it is people who choose to follow him.

What does a leader need to be like so that people follow him? Above all, he needs to be consistent in his thoughts, words and acts. What he says must coincide with what he does, in this order and in every possible way. As regards the safety realm, it is very common to see at plants leaders preaching that we must look after ourselves and that we have to carry out our tasks safely. But when the objective of a certain number of boxes or production or promised trucks is not reached, safety moves to a second position and "anything would do" to reach that number.

How to become a safety leader

If we refer to leadership in safety we have to follow or learn a series of steps. The first one is self-discipline. Discipline to change our beliefs and to start living safety as a value. And as such, it is non-negotiable.

Among the values which define a safety leader, self-discipline must be considered a value in itself and as such it must be put into practice and never negotiated. After all what we are talking about is the protection of the integrity and life of human beings

Here we can clearly see that it is not the same to talk about priority or values. Priorities change and are modified by external circumstances or even by different contexts or business. On the contrary, values remain immutable over our lives. It is here where the real challenge for a leader is: he must change his own beliefs, modify his acts, his mindset and his view of the world.

I summarize below the steps to follow to become Safety leaders:

- **Discipline.** It is evidenced by complying with the rules.
- **Perseverance.** It is established with patience and persistence.
- **Understanding.** Becoming fully aware of safety value. Understanding that it is at the base of empathy.
- **Rituals and habits.** They offer consistency in acts and every day life

- **Resilience.** It is the capacity to recover after unwanted results or adversity.

These steps must turn into transverse competences crossing every activity carried out by workers.

Organizational Change and its Developing Phases

When safety culture in an organization is referred to the focus is placed on the degree of maturity in this area. But in practice it is impossible to separate safety from organizational culture. An example: it is not technically possible for an organization to be in the reactive phase in the safety area and at the same time to be at the interdependent phase in the area of team work (always basing on Bradley´s curve).

For this reason and because it is more natural to start by changing the organizational culture through personnel care, safety culture is "separated" from the rest of the ingredients of the organizational culture and becomes a separate entity. Who can be against taking care of human life and improving working conditions?

The challenge here is for organization leaders to understand that protecting the physical integrity of its workers is part of their responsibilities and to acknowledge that it has to be considered an intrinsic value in business operations.

DC LANFRANCHI

Kairos Moment

We can describe the Kairos Moment as a single and unrepeatable moment which if missed, a unique opportunity would be lost. According to Wikipedia definition, Kairos is a concept from Greek Philosophy that represents an undetermined interval when something important happens. Its literal meaning is "right or opportune moment".

This might be the Kairos moment for a business, for learning, for a phase or for any important event in our career.

I will explain this with an experience.

In a multinational company a process of cultural change was taking place in order to establish safety as a value. One day at a meeting with managers, we were discussing the activities we had to carry out to demonstrate the operators that we were concerned about their safety.

The plant manager commented that on his way to the production manager's office, he saw an electrician on a chair fixing a curtain. He said that he thought to himself: "when I go back to my office I will tell him that it is risky to perform this task standing on a chair and that he should find a ladder".

The manager had the right thought, he identified the risk, but the urgency to go to the meeting was greater than his decision to stop this risky activity. When he went back his steps, the electrician was no longer there standing on a chair.

> *This moment represents the Kairos moment. This instant in which the manager missed the opportunity to genuinely worry about his personnel and to demonstrate his real commitment with safety by setting an example. And at this very moment the electrician also missed the opportunity to feel that his superior was concerned about him. The company missed the opportunity to show its point of view on safe behavior.*

Common sense

Excluding legal and highly technical matters related to special tasks and/or risks to implement safety is very simple. It takes basically, common sense.

It is habitual that in presence of a risky task, the chief of the division calls the safety expert to decide if it is possible to continue with a task or not.

It is here where we have to start using some words which apparently mean the same, but which in fact do not refer to the same thing at all. These words are: Accountability, Responsibility and Governance

Responsibility: It makes reference to the technician expert in a particular field, who has the responsibility to provide technical advice on everything related to his area of expertise (Expert- subject matter expert)

Accountability: It refers to your immediate senior, who has the non-delegable responsibility to make decisions necessary to protect workers, as he is the one in charge of accounting for any accident or risky circumstance (responsibility for everybody´s safety).

Governance: It makes reference to the one who has the responsibility to carry out an action. Here what is being delegated is the performance of the action, but never the responsibility for the control, mitigation/elimination of risks because responsibility is non-delegable (Superior).

While a task is being carried out and, according to the inexperienced view of an individual not specialized in the matter in question, the basic question to answer is: Would I do what this individual is doing? Would I let my son perform this task and in this way? If the answer is no, then your employee can neither do it ... Common sense.

PART 2 TAKING ACTION

Action Axes

The action axes suggested to manage teams are: Thinking, Taking action and Deciding. I will call them TAD.

Generally, action originates after thinking is completed. Thinking means that I have already elaborated on, reasoned, got trained and become aware. We do not act without thinking. We think before working.

Planning activities must include full knowledge and for this matter we suggest leaders emphasize effective communication and training.

In change management practices, the main axes are to communicate and train people. In order to make change sustainable, in the first place we must ensure clear and accurate communication supported by the right and appropriate training of abilities and knowledge. Secondly, this communication and training must be constant to turn it into a habit.

Communication

"I must understand first, and then be understood", according to one of the seven habits required to be efficient by Stephen Covey.

The way to develop this understanding is by means of an efficient and daily communication. It is not enough to have x number of boards with rules and news, it is necessary to develop communication habits.

Formal communication is done via boards, signaling, or any other device where everybody can write. Informal communications take place at meetings to follow up activities or at leisure times such as lunch at canteens and cafeterias.

Informal communications originate in the relationship of organization members, generally in groups with affinity and include casual conversation inside and outside the workplace.

In general, what happens is that communications do not possess the due implementation strategy and consequently they are not maintained through time. If communication is not kept active, reciprocal, easy, effective and fluid, the repetition of the message results in distortion and communications stop being important and turn merely informative and people stop paying attention to them.

If a board does not change periodically, it becomes unattractive. It is as if we were always watching the same news report, it becomes tiring and discouraging. This makes formal communications to be disregarded and depend exclusively on the leader´s charisma.

Even in disagreement, it is fundamental to achieve good communications. The correct use of the language encourages us to use adjectives and nouns correctly. To speak properly is essential to avoid misunderstandings.

Keys to effective communication

Communication problems are not only present in the business world. Nowadays, it is a real challenge to achieve successful communication everywhere and in every area of our lives. It is fundamental to develop good communication. Bad, poor or inefficient communication leads to misunderstanding. We know that misunderstanding may be harmful but is part of everyday life. The huge difference between misunderstanding in real life and misunderstanding in safety is that in HSE it causes accidents and deaths.

Consequently, what is basically important to achieve good communication requires it to:

- **Be accurate.** Use numbers, do not say words such as *"much"*, *"many"*, *"little"*, *"few"* or *"sometimes"*. Never say *"everybody gets late"*. Instead you can say *"1 out of 15 employees gets to work 15 minutes late"*.
- **Add value to the conversation.** Avoid being superficial, explain. Instead of saying *"the machine at the back"*, say *"the third machine located at the back of the area"*. Make an effort and do not take for granted that people understood your message.
- **Listen, observe and read through the others.** Pay attention to written, verbal and body language. The employee´s voice may suggest that the problem is not serious, but his eyes might be tough and show confusion.

- **Watch your posture, look into the eyes.** A typical mistake is to divide your attention among several things while somebody is talking to you, and this is not only disrespectful but causes insecurity to the person speaking as well.
- **Write down requests.** Do not trust your memory, take down notes, communicate in writing. Words count for nothing as the common saying goes. Avoid confusing directions.
- **Be kind.** We never know what the other person might be going through. From something trivial to an important personal conflict can distract a person from his responsibilities. Kindness generates empathy, trust and respect.
- **Never underestimate the volume of your voice.** A loud voice might be intimidating and a low voice might indicate a lack of interest in the conversation.
- **Avoid viruses "uhh", "ahh", "unm".** Instead, use "I see", "I understand", "that's true", "ok". The vague use of language produces emotional noise in communication because this vagueness can be interpreted as a lack of interest which in turn leads to misunderstanding.
- **Breathe calmly while speaking.** Agitated, short or irregular breathing is unhealthy because it contributes to stress.
- **Keep a good communication board:** Boards are useful, they provide important information which has to be constantly reviewed and updated.

Research by Susan Berkeley, worldwide referent in the development of messages through the voice, indicates that when we talk with a flat voice, with a lack of intonation or distracted, the effectiveness of communication is reduced to a 38% (yes, thirty eight per cent!). Imagine the impact of the use of a voice with such characteristics in the world of finance. We could assure it would bring about a very bad scenario. **Now, imagine a reduction of a 38% in communication effectiveness in safety matters.** Completely at a loss for words.

Managers underestimate the impact everyday matters such as verbal communication has on business. I wonder if they would feel the same supposing this happened within their homes, with their partners, even with their pets care

Communication and Cultural Differences

Taking into account cultural differences is a critical factor leading to success. For this matter, it is essential to rely on a model of cultural considerations.

This helps us predict how people from different cultures will speak, behave, negotiate and make decisions.

Communication and Cognitive styles

Abstract Thinkers vs. Associative Thinkers

An environment where problem solving is taught develops abstract thinking. Abstract thinkers can deal with something genuinely "new". They are individuals who due to their characteristics are able to deal easily with critical situations. On the other hand, associative thinkers generally understand situations through the screen of personal experience, they are only effective in critical situations if they had gone through similar situations.

Particularistic vs Generalistic

We already know the abstract model versus the associative model. But we also need to differentiate whether they are particularistics/personalistics or generalistics/universalistics. Particularistics feel that a personal relationship or situation is more important than obeying the rules and regulations. In contrast, universalistics tend to observe rules and regulations, proving that general welfare is above their personal situation. We can see this, for example, in people who prefer not to argue in public with people they like (particularistics) and others who consider a natural behavior to talk about facts and situations without any personal approach (universalistics).

Negotiating Strategies

Faith and Evidence

To negotiate, do we have to stipulate in advance what we accept as evidence? To clarify, we need to see the difference between facts and judgment.

What is truth? Different cultural styles perceive the truth in different ways. These ways can be summarized with three words: faith, facts and feelings.

People can have faith, beliefs rooted in human beings. The truth depends then more on the faith we have in the matter.

We can make decisions depending on facts, objective data deprived of personal data.

Or we can act according to our feelings. Let´s keep in mind that feelings boost intuition and that intuition is the mixture of knowledge and experience.

Independently from our negotiation models, what is really important as far as safety is concerned, is to synchronize the styles of working teams. When a manager puts together somebody who is feelings driven with people who rely on data we can observe that when they debate they share the same opinion but from a different perspective.

Systems of Values

The author Terri Morrison, world famous expert on the analysis of cultural models, emphasizes the importance of analyzing the base of values behavior in our group.

There are 3 systems of values:

Inefficiency in decision making: there is generally no coherence between individualism and collectivism.

Sources of anxiety: a combination of relationships, religion, technology and rules or regulations are the safety and stability sources people need to deal with stress.

Equality and Inequality: Every culture has disadvantaged groups, we will focus on certain sectors with inequality.

Mathematically, accidents have a direct relation with these three groups of values. There follows a comparison of an American model (US) with a South American model.

> *In South America people are more concerned about the consequences of an activity than about the activity itself. While in the United States, they are highly analytic, facts are important and the rules of an organization prevail over those who negotiate. One model is associative, the other is abstract and focused on problem solving.*
>
> *In South America negotiation is acceptable as long as feelings or rules are not contradicted. It is very difficult for a South American to argue with a friend. Decisions tend to include consensus with friends and acquaintances. On the other hand, in the United States negotiations take place while gathering objective facts and circumstances.*
>
> *Which is the cultural model of the organization? Which is the cultural model of the leader, his school, his inner circle?*

Where and how to take action?

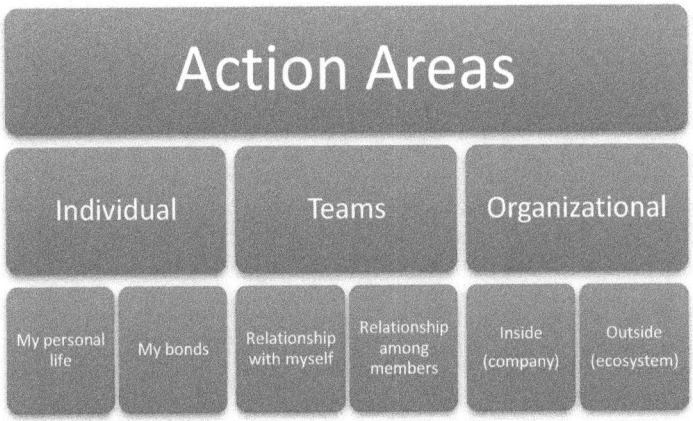

Action areas represent the scenario where decisions are made:

If I work in an individualistic way the considerations and perspectives of teams and the organization´s rules lose balance. The same is true if I prioritize any of the three component parts. This graph is not included by chance. I want you to note that the parts allocated to each level are identical and that each level is symmetrically identical to its parts.

PART 3 MEASURING

The importance of combined variables

In the first part of this book, when I was writing about coherence, I explained the importance of studying working metrics in depth.

It is not only necessary to elaborate bar or column graphs or comparative areas and trends, but to combine variables with the results of other variables. There is where the secret of depth analysis lies. Each time combined variables are used I can appreciate metrics in depth. These metrics are the ones that offer real value at decision making.

A company has employees and employees´ payroll varies every day, so it is necessary to estimate daily production over the total number of employees of this day. It is easy to unmask the effort of a prize for exceeding production if we increase production with extra working hours. In turn, employees must be measured in human hours and machine hours. Each company must have its own production rolls which it will use to suit its needs.

What is of interest, being the theme of this book, is human safety, and so what needs to be analyzed is human hours and not machine hours. These human hours can be the addition of shifts, as averages. Human effort must be measured in comparison with 2 areas: business and human life and it is there where we find a balance of coherence formulas.

Some years ago, I was told that in a vice presidents meeting in a multinational company rated within the 20 biggest companies in the world, one participant challenged all those present to contradict his belief that it was not possible to elaborate formulas that guarantee efficiency or safety. He supported his view stating that each plant, each area, each city and each country was different, and thus as they operated in more than 100 countries, to develop behavioral reports was a complete waste of time.

The person who told me this was very annoyed and he accepted the challenge and months later submitted reports which deserved an international distinction. He managed to compare 19 countries implementing behavioral metrics.

What was outstanding about his task was the comparison of some metrics of more than 300,000 employees taking samples of the best evaluated population of the organization (top performers, president´s club, distinguished engineers, people doing patents, prizes and bonuses) and it was combined with 2 elements: the geographical location of these people and the production and /or economic results of the areas to which they belonged. It was found that in every "neighborhood" where there was people recognized by their seniors, the best efficiency parameters were observed. This is common sense, but demonstrated in Excel.

This leads us to say that there is further information an Excel can provide us with, and that due to its complexity, it would escape common sense.

It is vital for a leader to have this information to achieve good performance.

Even if all the information provided by the metrics of combined variables is useful to a certain extent, after many years of using metrics, I have learnt two fundamental rules: one of the rules states that comparing progress is more useful than comparing results and the second is the Rule of Seniority of Reports.

Comparing Progress

There is no formula which is homogenous for all teams and scenarios, but it is possible to compare each plant, each team, each country with their own metrics of previous periods. In this way, it is possible to elaborate comparable progress ratios which being already deprived of specific data, can in fact be compared.

In other words, there is no formula to assess how effective we are or how safely we work, but we can perform some calculations to see whether we improve or not.

Briefly, if we move one millimeter ahead in our task daily, we will have moved forward three meters, sixty five centimeters per year towards our objective.

Seniority of Reports

The golden rule I learnt about reports is: **Choose the most qualified person to elaborate reports.**

Reports represent the base on which executives rely when making decisions, and because of this we cannot leave this task in the hands of junior staff. If this happened, we would end up obtaining just figures, not valuable information.

Let´s see the most widely used profiles:

Financial Investment Analysts: they know about finance and profits and losses (P/L), but they are not experts in macros programming, data mining and even less in data base management.

Engineers: Generally the staff chosen has little experience in business and most of the time they use complicated fomulas and complex tools which are extremely difficult and expensive. Engineers with great human abilities are the ones prepared to provide solutions.

Psychologists, sociologists and human resources specialists: They are the best qualified to understand some metrics and spot problems, but their poor knowledge of spreadsheets and data mining tools makes it impossible for them to elaborate reports.

Do not then look for profiles that conform to stereotypes. We´d better focus on the abilities we need. I am convinced that in the company there are many managers who possess these abilities and we can invite them to elaborate reports.

Let´s do a list of the seniority we are looking for:

1. Managerial profile with many years of experience, not less than 15 to 20 years.
2. Expertise in the use of Microsoft Excel, dynamic tables and macros programming.

3. Ability and empathy to work with IT areas to elaborate macros to see data bases and empathy with human resources staff as well to have access to permits and confidential information of the company.

Why do we consider the spreadsheet so important? Because every director and executive in an organization has office tools such as word processors, spreadsheets and power point. Most of them can read Microsoft documents and spreadsheets. But if we give them a complicated application with permits they would not use it.

This is an experience a client shared with me.

An executive of a multinational company was relocated to London responsible for the operation of over 100,000 employees in Europe and Asia.

In Europe one management system was used and in Asia, as the result of several mergers, other systems were used. Some used SAP, others Oracle, and others their own systems.

Even if this executive had access to many systems which were in turn connected with thousands of information subsystems, he could not construct the information he wanted to compare.

The solution was to ask for Excel reports, and with the help of a manager they prepared macros which consulted huge quantities of files, gathering data in working sheets with dynamic tables. This was the information this executive needed.

The inclusion of trend lines which crossed systems that had never been compared before, allowed the organization to anticipate and make decisions which

> *generated many million dollar profits. These profits were used partly to develop safety systems in order to reduce accidents.*
>
> *Success undoubtedly was achieved by choosing experienced people to elaborate reports.*

We have seen that accidents are foreseeable if we count with a good data analysis. The requirement is to involve people highly trained in the comparison of data about time of arrival to the workplace with weather data, production and quality data. Accidents do not occur by chance. They are the consequence of situations which should be studied and foreseen.

It is necessary to do research, encourage the use of spreadsheets and rely on external consultants, who even if extremely expensive, provide us with suggestions that will result in savings and profits thanks to the decrease of fatal incidents and accidents and higher efficiency rates.

REPORTS = SENIOR ABILITIES

Detection of Mathematical Patterns

Safety requires the creation of mathematical patterns.

What is a mathematical pattern? It is a model which if studied periodically offers similarities or patterns.

For example, the study of variables of accidents on rainy days versus on sunny days. Clearly, the occurrence of road accidents is always more related to weather conditions

Mathematical patterns result from the study of parameters over time, multidimensional. In Coca Cola Company 30 variables were analyzed for some studies while for the control of credit card fraud more than 120 variables might be compared.

Consequently, to study mathematical patterns we will need 2 elements. Firstly, Excel tables which will show in their main columns data related to time as dates, times, months, weeks, seasons (summer, autumn, winter, spring) and any other information about time. The columns that follow will contain every attribute to be analyzed (people, geographical locations, accidents, kinds of accidents, weather conditions and the tens or hundreds of variables that impact on an organization).

After this, we have to choose pairs of data. Patterns are always studied by pairs of data. To this matter, we will utilize tools for data network analysis.

To make a graph of networks we will include the following categories:

- **Networks 1.0:** we associate the values to a single node.
- **Networks 1.5:** we also consider the connections among the elements linked to the node under study.

- **Networks 2.0:** we study how variables are connected to other nodes under study.
- **Clustered Networks:** we remove the node under study and we observe the behavior of the variables studied in networks 1.0 and 1.5

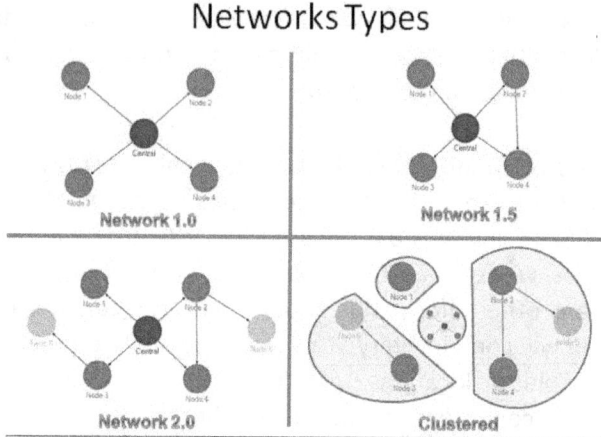

Below we can observe a data network which illustrates how the elements group in a cluster. This graph shows clusters by country in a multinational company.

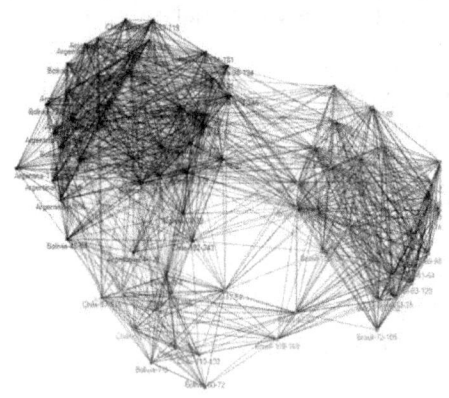

Which is our safety challenge? To go on doing research until we conclude that accidents are grouped in the same group of variables. It is there where patterns become visible. And thanks to this we are on the right track to find solutions.

In the following example we can observe the same information about accidents, but grouped in a different way. Like this, we are finding patterns for accident behavior which will enable us to predict them and avoid them.

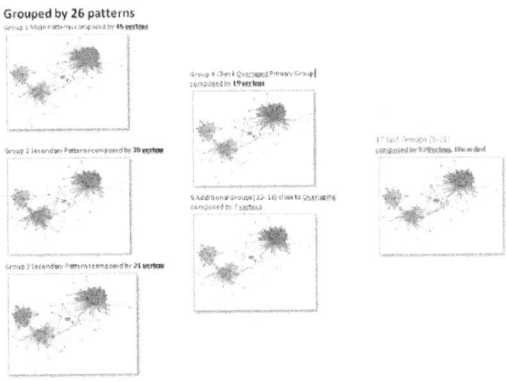

How important is it for a leader to measure?

As we have said that measuring is the most important action in an organization because it is the quantification of data, it is very important for a leader to measure since these data provide him with the information necessary to manage results in his team and to administrate safety appropriately.

To clarify, we cannot manage what we do not measure. What is not measured, badly measured and/or misunderstood is badly managed.

I invite you to have periodical meetings of Review, Action and Measurement. To compare the previous variables with the present variables will show you the progress rate of your performance. Without measurements there is no progress. With measurements we create a virtuous circle where we anticipate, take action and measure and so on.

PART 4 AFTERTHOUGHTS & REFLEXIONS

A crisis occurs when the old does not finish dying and when the new is not yet born

Berlot Brecht

Why do crises happen and the end? April 2020 Reflections

All crises happen and then they go away because everything does, and all that remains is what we were able to build in both times of peace and in times **of turbulence..**

When we talk about issues of Safety it does not escape that law. Today more than ever, as leaders, we must be alert and vigilant about the health and safety of our people. And when I say our people, I mean our family, friends, work teams, neighbors, collaborators and ourselves.

Self-leadership in terms of safety has become, in this time of pandemic, an even more fundamental issue than it already was before. To not perceive the risk that surrounds us today without the seriousness and social consciousness necessary is a trap in which we can fall from considering the enemy an "invisible" one, as are many of the risks that surround us on our daily life. And that's why we let our guard down, because we don't see it or we get used to living with it, or we create to ourselves a false sense of safety where we make ourselves believe that things are ok, things are going well or things are getting better on their own.

And so, we allow ourselves to loosen up our defenses and common sense in the way of building the awareness of safety and health. That´s a journey that does not have a happy ending.

To achieve some sort of happy ending in post of the caring of the most precious asset that we have, which is our life, we must work permanently in the search for continuous improvement.

Improving our organizational processes, improving our safety management processes, improving the safety culture and above all things improving our own skills as leaders and even more so as people, it is critical to be able to face the challenges of this new world that is brewing. It is true that every crisis brings new opportunities in return, but in order to take them you must be prepared.

We have in our hands the opportunity and responsibility to exercise our **visible leadership** in front of our people. We have the non-delegable responsibility for the caring of others that we implicitly accept when we claim to be their leader, their guide. The commitment then is to get into action. Let's do it!

Today more than ever our people need committed leaders, with courage and humanitarian skills and heart, ensuring the safety and health of their teams. Let's seize this moment. What we are living today is unique and historic. Let us work up to the circumstances and the needs of our people.

Many companies have the BCP -Business Continuity Plan- with crisis committees and emergency preparations. Such committees formulate simulations and tests of abnormal situations but under normal conditions. However, for the times that we live nowadays no one was prepared, and it is even more complex, since most people had not even thought about it.

Well, let me rephrase that last bit. The truth is that yes, many people in universities and government agencies thought about a scenario as the one we are living today. It was also thought by many creators of science fiction movies, but as something that can only happen in our imagination. Some lateral thinkers like Bill Gates talked about it in 2015, but he was taken as an exaggerated idea or very advanced in time.

And yet here we are, starting the year 2020 in which the worst pandemic in history has been declared. A microscopic entity, a virus, a Coronavirus, that scientists are still discussing whether a virus is a living being or not. And in the midst of all that discussion the virus is killing us and is killing the world, literally and not so literally speaking.

The worst deaths are the real ones, the deaths of thousands of real people, no doubt about it. However, before the situation normalizes, this global crisis will shake us with its aftershocks, as all great earthquakes do, that will be many and there will be plenty of other types of deaths.

Many things and dynamics will definitely change. Money and the relationship with money will change. Perhaps cryptocurrencies and virtual money will be kicked out of the investment portfolios of a few enlightened ones and get pushed into everyone's daily life in the rise of the world of the new order. That would prevent, for example, queues at banks, pharmacies and supermarkets.

Major and important changes in cleaning and hygiene rituals, both in businesses and homes, are changing and that change has come to stay.

And without a doubt what has changed forever is work. Everyone's work. Many of us resisted the virtual world, because the one-on-one contact was part of what we wanted to feel, we wanted to feel close to other people. Work will forever change as will many other experiences that we believed it could only be done by being physically present. Today´s reality confronts us with many of our limiting beliefs, many of which we live and experience in so many areas of our lives, and that are also to be found the area of safety.

The year 2020 came and from one day to the next, suddenly and in a blink of an eye, without asking permission or giving us time to organize anything, our home became our office. But not only our house was transformed into the office of all those who work in each home, but it was also transformed into a school, a kindergarten, a gym, a restaurant, a hair salon, everything!!! Our homes have been transformed into everything 7 days a week 24 hours a day. Many of us were brought a computer to our house the very next day and our children were no longer welcomed at their schools. At that very moment the real change began. Because in addition it was no longer allowed, nor was it safe, to leave our houses.

To work well we must have a space that accommodates all our needs and requirements of all kinds. That is why it is extremely important to have a safe, healthy and happy design and equipment for the home office workspace.

Let's look at several important points that we can contribute from the Safety area for this new stage of a home office:

Ergonomics

Ergonomics is the science that adapts the job to the activity and the person who develops it and not the other way around. Since the industrial revolution was always prioritized or played with the incredible adaptability that human beings have to everything.. Even viruses!! And then tasks arose at very high temperatures, with heavy materials, with forced positions based on that imposed carrot, sometimes wonderful, that drives us to believe that human beings can do everything.

However, revolutionary thinkers emerged and with them the concept of Corporate Social Responsibility (CSR), which made us see and understand that not everything should be done, even if possible.

Although the term was first coined in 1857, in Poland, it was in the 1970s, well into the 20th century, that ergonomics took its relevance as we know it today.

The objectives of ergonomics are:

- To reduce or eliminate occupational risks, labor accidents and diseases.
- To decrease physical, psychophysical and mental fatigue
- To increasing the efficiency of productive activities

The bases and laws of ergonomy are already established and tested, and by them it is understood that for the safe, healthy and happy design of the workplace, the following four principles must be taken into account:

1. **The locative risks**: i.e. the risks inherent in the workplace itself.

- Detect loose objects
- Don't stan on tables or chairs to reach objects
- Close drawers to avoid bumps and stumbles, among others possible accidents.

2. **Ergonomic risks:** are those that arise when we neglect our bodies position during the execution of our work.

- Take into consideration the importance of lighting, the more natural and direct, the better. Avoid irritating and uncomfortable reflections on work surfaces.
- Arrange the more frequently used objects nearby and at your fingertips, and the least frequently used objects in a wider radius distance.
- Keep your elbows and knees always at 90 degrees in relationship to the body axis.
- Keep your feet firmly resting on the ground. To accomplish this, you can use shoe boxes, wooden drawers or cushions.
- Keep the dorsal spine comfortably supported by the back of the chair (yes, you have to choose a chair with backrest), if you can´t get your back to rest comfortably on the backrest you should try using some cushions.

3. **Electrical risks:** these are more common than we think and perhaps the least measured.
    - To avoid them keep the cables neat and out of the way to avoid tripping and falling over
    - Don't place glasses with water or infusions in places where you might

inadvertently dump them on your keyboard or computer
- Don´t overload plugs
- Don´t use extenders that are not approved by a licensed electrician
- Electrical installation, both at work, and at home must have differential protection and circuit breakers

4. **Psychosocial risks:** These have become more relevant today than ever before due to periods of obligatory social isolation, but it goes without saying that home office work has always been and always will be a challenge, especially regarding the maintenance of fluid and healthy interpersonal relationships, even more so during these special times.

Building happiness

In addition to everything mentioned above, it is important to have a work environment that makes us happy, that provides us with well-being, comfort and joy. And so, we must build that environment. Yes, build it. Because the environment that leads to happiness is built in all areas of life.

To build happiness we will use the science of positive psychology. This science ensures that 50% of the level of our happiness is due to our hereditary base, 10% is due to the circumstances and the remaining 40% is due to the intentional activities we do in our lives. That is why we can affirm that happiness is built, it is a choice that we can make. Science shows that we have a 40% of "space" to shape our lives with nice thoughts and feelings that come from doing things we like and makes us feel good.

Healthy Routines

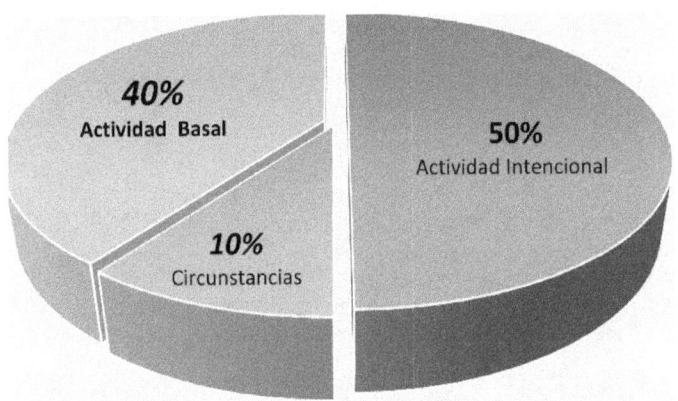

Another important part of our home office routine is being able to make **active micro breaks.**

Active micro active breaks or micro active pauses serve us to make some muscle stretching exercises. It is necessary to take them every certain interval of time to help prevent diseases typical of those who spend many hours in little or no movement and can expose us to two specific dangerous risks: sedentary lifestyle and non-ergonomic body posture.

Unlike an accident, which is a sudden and unexpected event, occupational diseases manifest over time, as they are the consequence of cumulative and gradual exposure to poor work habits.

> Obviously, for an occupational disease to develop, an individual must also have a basal load of predisposition. There are people who, at the same stimulus, do not develop a disease that others do, or develop it later, or it manifests itself with less severity.

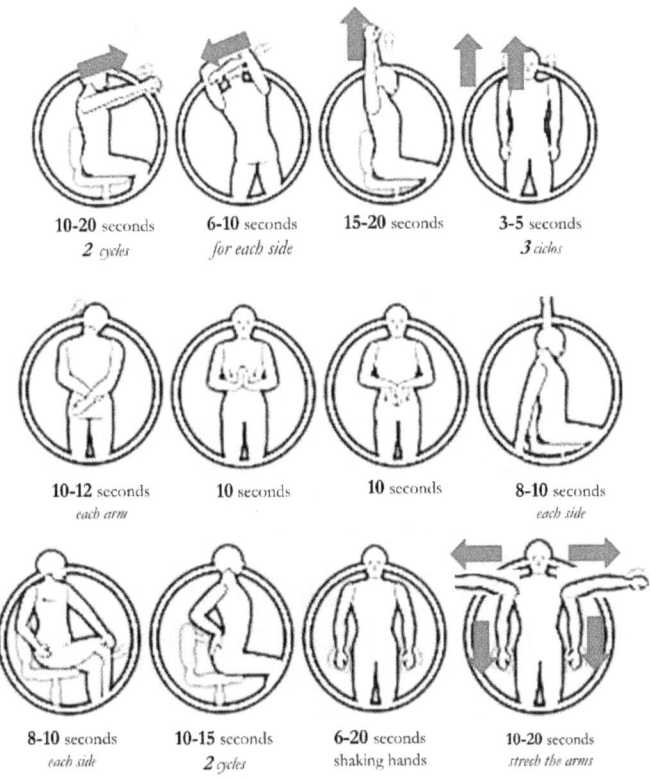

| 10-20 seconds | 6-10 seconds | 15-20 seconds | 3-5 seconds |
| *2 cycles* | *for each side* | | *3 ciclos* |

| 10-12 seconds | 10 seconds | 10 seconds | 8-10 seconds |
| *each arm* | | | *each side* |

| 8-10 seconds | 10-15 seconds | 6-20 seconds | 10-20 seconds |
| *each side* | *2 cycles* | shaking hands | *strech the arms* |

Performing 3 to 5 minutes of micro-breaks every three to four hours of work will improve not only our body, but also our attention, our mood and our mood. Staying on the move is extremely important to our psychophysical balance.

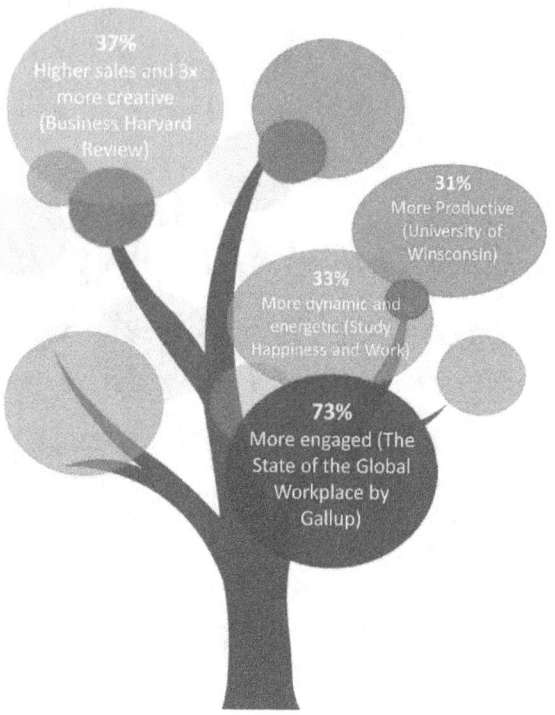

- 37% Higher sales and 3x more creative (Business Harvard Review)
- 31% More Productive (University of Winsconsin)
- 33% More dynamic and energetic (Study Happiness and Work)
- 73% More engaged (The State of the Global Workplace by Gallup)

Happy companies are those that make sure their employees are happy, because organizations are the sum of all their employees.

Human beings are a unique and indivisible entity, and from ontological coherence we are what we say, and that transforms us into what we do. To be happy we must think, feel, say and do in a coherent way, that is, harmoniously.

It´s a good thing that many years ago we got rid of a very limiting belief that said that personal problems could be left at the entrance door of the office, or at the entrance gate of the factory, or in the entrance hall of the house next to the umbrella and the coat. Today we know that this is not the case anymore because it´s neither possible nor healthy.

My reflection in this annex is purely to invite you to walk this path of coherence and allow emotions to emerge in the area that needs to emerge, thus avoiding much stress.

Stress

The workload that the employee believes they cannot handle is called job stress, and nowadays it is the second most important and frequent cause of work sick leave followed by musculoskeletal causes.

What is striking is that both causes, musculoskeletal and emotional, are affecting all industries and organizations without distinction of which category they belong to.

Today, in the midst of the Coronavirus pandemic spreading through our streets, the two most prevalent causes of sick leave are in full bloom.

We have the feeling that our bodies and minds have been imprisoned by an invisible enemy against which we cannot fight. But it's not like that. We can fight against it from what we know, which is to take care of our body with the tools that give us safety, hygiene, ergonomics and above all our minds, with the tools of positive psychology, creativity and mental flexibility.

.

Happy and healthy employees build happy and healthy organizations. Healthy organizations have been proven to be more efficient and profitable business. Therefore, it will be the organizations that understand this important concept and invest on it all their resources (human, capital, know-how, lateral thinking, etc.) the ones that will survive to these or any other time of crisis.

EPILOGUE

I wrote in the introduction that safety has always been a passion in my life. And when I write it, I do understand how deep and broad this concept is to me. It goes far beyond corporate safety. My passion comprehends everything about feeling protected because feeling protected is fundamental in our life.

It is for this reason that working as a consultant and giving workshops to enhance safe environments constitutes my utmost commitment, both with a worker and his whole family. Mi mission is to ensure that all of us work safely and go back home with our families safe and sound and enjoy life.

Writing brief and easy reading books as a guide to safety training is the means to achieve my objective of offering safety methodologies while re humanizing us at the same time.

I do not want to add more words, I just want you to press the SAFE button and I want you to remember when you meet your relatives and friends how good it fees to be protected and safe. Safety is the basis of a good and productive life.

If you wish to share experiences, information or suggestions, please write to me at debbie@resiliere.com. I cannot promise I will answer all the emails, but I do promise I will read all and each of them. I am deeply grateful for your contribution.

And remember. One day the life of a worker will be in the hands of somebody else`s decisions. I hope with all my heart, that all of us have contributed, as a team, so that the decisions taken are the right decisions.

Deborah C. Lanfranchi

Experienced Leader in HSE and Manufacturing; with vast working experience in Multinational Industries. Expert in Corporate Management Systems, Risk Management and ISO Standards. Certified Professional Coach, Life Coaching. Graduated Chemical Engineer at UTN (Public University of Technology) and UBA (University of Buenos Aires); associated to the CPIQ (Professional Registry of Chemical Engineers).

Education

Chemical Engineering. Universidad Tecnológica Nacional – (Public University of Technology) Argentina.

Professional Ontological Coaching.

Specialized in Industrial Safety and Hygiene – Universidad de Buenos Aires (University of Buenos Aires) – Argentina.

Teacher at the Post-graduate course in Safety and Hygiene Specialization, School of Exact and Natural Sciences - Universidad de Buenos Aires (University of Buenos Aires) - Argentina.

NOTES

NOTES

NOTES

NOTES

NOTES

NOTES

www.ingramcontent.com/pod-product-compliance
Lightning Source LLC
Chambersburg PA
CBHW070802220526
45466CB00002B/516